BATTLE DISCOVERY
‹2›

THE BATTLE OF CANNAE

216 BCE

PATRICK HUNT

STONE TOWER PRESS

The Battle of Cannae 216 BCE

Stone Tower Press
7 Ellen Rd.
Middletown, RI 02842
stonetowerpress.com

Contents of this book appeared earlier as a chapter in the author's *When Empires Clash*, also available from Stone Tower Press. The present work is revised and updated.

ISBN: 979-8-9919831-9-8

Formatting and cover design by Amy Cole, JPL Design Solutions

Editing and proofing assistance by Jacob Alexander

Maps by A. D. Riddle, RiddleMaps.com

Front cover illustration: Classical monument to the memory of the fallen at the site of the Battle of Cannae, photo and copyright by Jörg Schulz, used with permission.

Printed in the United States of America

TABLES OF CONTENTS

PREFACE

"The predominance of the psychological over the physical, and its greater constancy, point to the conclusion that any theory of war should be as broad as possible. An intensive study of one campaign unless based on an extensive knowledge of the whole history of war is likely to lead us into pitfalls ... Effective results in war have rarely been attained unless the approach has had such indirectness as to ensure the opponent's unreadiness to meet it."

Sir B. H. Liddell Hart, *Strategy*, 25

HOW OFTEN HAS HISTORY pivoted around one great battle or siege? When empires have clashed, how often have outcomes been surprises—a lack of readiness as Liddell Hart mentions above—or contrary to the expectations of a military opponent? Not all ancient battles or sieges changed history forever, but some stand out dramatically for multiple reasons. Whether for unusual tactics, unexpected outcomes, bold and brilliant command, quick adaptation to the changing field circumstances, prescient awareness of the "fog of war" or some other factor, some great battles changed the playing field forever, either bending the destiny of perceived military juggernauts and shocking the world in which they happened or forcing cultures to redraw the rules of engagement. The legacy of such engagements is enormous, often lasting centuries. In the pages that follow is the story of one such event.

THE BATTLE OF CANNAE

216 BCE

"A vast mass of 70,000 brave souls was encircled in a small plain in southwestern Italy by a poorly organized but brilliantly led invading army of half its size. Confusion and terror only grew greater when dusk neared, as each Roman pushed blindly and was shoved into the enemy on all sides. Stacked in rows of thirty-five and more, the size of the unwieldy mass began to ensure its destruction."

Victor Davis Hansen,
Carnage and Culture, 99–100.

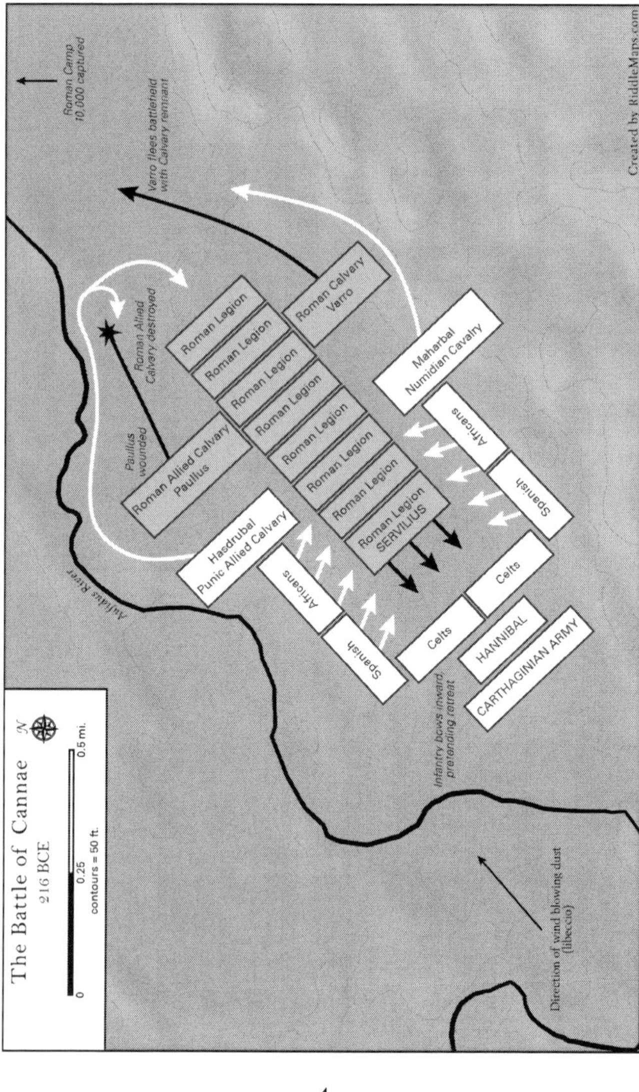
The Battle of Cannae
216 BCE
N
0
0.25
0.5 mi.
contours = 50 ft.
Roman Camp
10,000 captured
Varro flees battlefield
with Calvary remnant
Roman Calvary
Varro
Roman Legion
Roman Legion
Roman Legion
Roman Legion
Roman Legion
Roman Legion
Roman Legion
Roman Legion
SERVILIUS
Roman Allied
Calvary destroyed
Paullus
wounded
Roman Allied Calvary
Paullus
Maharbal
Numidian Cavalry
Africans
Spanish
Hasdrubal
Punic Allied Calvary
Africans
Spanish
Aufidus River
Celts
Celts
HANNIBAL
CARTHAGINIAN ARMY
Infantry bows inward
pretending retreat
Direction of wind blowing dust
(libeccio)
Created by RiddleMaps.com

BACKGROUND

HANNIBAL OBLITERATED WHATEVER ARMIES Rome threw at him in Italy in a lightning three-year succession of battles from 218–216 BCE, from Trebbia River (or Trebia River) to Trasimene to Cannae, each different in the showcasing of his tactical genius. What the Battle of Trebbia River (or Trebia River) began (218 BCE), and Trasimene continued (217 BCE), Cannae concluded (216 BCE), being considered Hannibal's greatest triumph, a paradigmatic example of his military leadership and apotheosis of decisive victory.[1]

Such a risk-filled invasion as that of Italy needed to be quick, not bogged down with long delays where the enemy at home has the luxury of replenishing its army and adapting to the type of warfare Hannibal brought. If one of Hannibal's primary aims was to separate Rome from its Italian allies, to persuasively cause a mass defection to his side would require a stunning and comprehensive Roman defeat.

For awhile, the desperate Romans had denied Hannibal this chance, replacing temporarily the dual consulship of military command by setting up old Quintus Fabius Maximus Verrucosus[2] as temporary dictator, whose strict policy of avoiding confrontation gave him the nickname of *Cunctator* or "Delayer." Fabius was a conservative military veteran patrician, a twice-elected consul in 233 BCE and 228 BCE who had the popular support of both his own class and among the common people whom his ancestors had championed against patricians. He had restored morale after the Roman debacle at Trasimene against Hannibal and had made sure Roman religious observance was placating all the national gods.[3] A frustrated Hannibal had to live off the land he had invaded with his army, fed by purloined supplies from an increasingly resentful Italy and harassed by Fabius's Roman raiding parties who would dodge direct confrontation.

This war of attrition was exactly what Hannibal did not want. He knew Rome could continue to raise endless legions against him while his own numbers dwindled gradually; as both his mercenaries and the Celts who needed loot to stay content in Italy would not be advantaged by waiting around for a battle that never came, marching from place to place across Italy, and setting up winter camp in increasingly hostile territory. Hannibal walked a narrow line between trying to keep his army intact and fed and having to take what he needed from a neutral or marginally loyal Italian population he hoped to win over.

Hannibal escaped being bottled up by Fabius in the abundant *Ager Falernus* (Volturnus) of Campania one night by using thousands of rustled cattle with firebrands tied to their horns as a surrogate army. This ruse made the Romans think his army was moving in an alternate direction and the befuddled Romans, seeing only moving torches in the dark, abandoned their guard in the low pass Hannibal needed to cross, so he made his way out of rich Campania to a safe winter camp eastward. He had counted on Fabius to stay put until morning rather than risk a big defeat, and Fabius did not disappoint.[4]

But now contrary to the mostly successful yet unpopular Fabian strategy of non-confrontation,[5] Rome raised eight more legions[6] (at least 80,000 men) by the late spring and summer of 216 BCE, and the Senate felt comfortable to take the fight to Hannibal again with confidence in the largest troop levy Italy had ever seen. The fuse was lit when Hannibal, always needing food and supplies, and particularly so with harvest over and autumn looming, quickly took the Roman grain supply depot near

Canusium in Apulia.[7] The mostly flat and broad plain near the Adriatic beckoned irresistibly to the now full Roman army, suggesting an optimal battle place for their eight legions so that about a week later the huge Roman assembled forces came within sight of Cannae, a *vicus* (village) of Canusium.

It is likely that Rome had spent so much time and energy in recruiting and assembling the new legions that they had little to no time for training them except on the march since the combined Roman armies were probably only together a week before battle.[8] Hannibal had not wasted any time and was more ready than the Romans could imagine, although Cannae was a risky field for battle that could advantage Rome unless he could control as many variables as possible. It was probably nearing the first day of August in 216 BCE when both armies camped a few miles apart and the battle loomed.

DISCUSSION OF TOPOGRAPHY

Cannae, in southeast Italy, is about 4.5 miles west of the present Adriatic coastal city of Barletta, lying between it and Canosa di Puglia (ancient *Canusium*), inland and south of the Gulf of Manfredonia and about 40 miles northwest of Bari. The stream rich in oxbows that bisects its low valley from southwest to northeast is known now as Ofanto and is derived from its ancient name of Aufidus. A low hill with the *vicus* ruins of Cannae is just south of the stream valley, now marked by a memorial column. From it, the whole battlefield presents itself below one's feet as the author has regarded it like so many others. Some debate remains about the exact location of the main battle and the enemy camps as the Ofanto stream has altered its course somewhat over the millennia because the eroded terrain is generally low except for the hill with its ruins.[9]

Many believe the battlefield was south of the Aufidus on the right bank if facing east to the sea.[10] If this is the truest site, much of the battlefield is only about 4,000 feet maximum or 0.8 of a mile wide between the bluff of the overlooking hill and the currently winding streambed; about treble that in length or 1.6 miles long depending on where forces were concentrated. Intersected by the Strada Provinciale 142 just west of the hill near the Chiesa San Nicola, the Strada Provinciale 3 dissects the plain mostly east to west with its adjacent railroad but is closest to the hill bluff with its *vicus* ruins and its Parco Archeologico aptly named "Canne della Battaglia." About 3.5 miles west of the battle site is the modern village of San Fernando di Puglia. On the other hand, Lancel has the battleground slightly to the east of the hill closer to the sea,[11] but even here the topography is still fairly constrained between the river and low hills.

The overall site of the battlefield is now covered with vineyards whose vines grow in a calcareous white soil. However, enriched by so much blood and bones so long ago, such normally pleasant vineyards in no way reduce the horror of what transpired that early August day of 216 BCE.[12] For this author the eerie presence of innumerable flies on a hot August afternoon was somehow reminiscent of the kind of carnage that would attract flies to breed there for a long time. I walked the overall battlefield length and breadth over and over and was astonished how small the low valley still is to have contained such a massive death toll. Some maintain the valley may be only slightly smaller now than then, with the river now flowing a little closer to the hill, but this is arguable.[13]

Hannibal's first campground seems to have been on the village hilltop of San Fernando di Puglia, whereas the first two Roman camps were likely a mile or more to the east[14] and probably south of the battlefield but mostly straddling the Aufidus.[15] Although arguable, the Romans may have then moved at least one of their camps away from the Aufidus because Hannibal seems to have been closer to the watercourse of the Aufidus stream, whose water he protected, making it difficult for Romans to access because of harassing Numidian calvary.[16]

MAJOR COMMANDERS

For the Romans, the two primary consul generals were Lucius Aemilius Paullus and Gaius Terentius Varro, opposites in almost every way. There were at least 200 senior Roman equestrian officers at Cannae, and likely up to 80 of senatorial rank as well as up to 60 tribunes, the army *quaestors*, and even the two immediately prior consuls Gnaeus Servilius Geminus and Marcus Attilius Regulus and a previous co-commandant of Fabius, the *magister militum* Minucius Rufus. The main veteran consul commander Aemilius Paullus was from a most distinguished Roman patrician family—one of the first families of Rome in the *gens Aemilia*—who had won a victory over the Illyrians in 219 BCE. But his reputation had been dented unfairly by a dubious scandal of unequal distribution of spoils. Terentius Varro, on the other hand, was a "new man" who had risen from modest means. His power had come mostly from the manipulation of commoners via demagoguery and accumulating plebeian favors. He had little to no military experience and was considered a madman by Fabius according to Plutarch, before Cannae:

> Fabius tried to rouse and encourage him [Paullus] to restrain the madness of his colleague, showing him that he must struggle to save his country not so much from Hannibal as from Terentius.[17]

This was needed because of the Roman practice of alternating daily command between military veterans and politicians and, Hannibal knew all too well how to exploit it.

Hannibal again led the Carthaginian forces as at Trebbia River and Trasimene and the Volturnus escapade; Maharbal was the commander of the Numidian light cavalry on the right wing and the

Carthaginian Hasdrubal commanded the left wing of allied cavalry. Hannibal's younger brother Mago was again likely present as at Trebbia River. Few other officers are named in Hannibal's army, except one Gisgo, who remarked how large the Roman army was in comparison to Hannibal's. Hannibal's legendary wry and perhaps even encouraging reply was, "Yes, but not one of them is named Gisgo."[18] Plutarch suggests that the laughter that rippled through the Punic ranks before the battle steadied the troops by reducing anxiety over visible Roman superiority in numbers, which would have been daunting to anyone, even Hannibal's battle-hardened veterans.

ORDER OF BATTLE

Rome brought its largest ever infantry and cavalry force to date to Cannae, probably at an infantry to cavalry ratio of 13:1 (at least 80,000 infantry and 6,000 cavalry),[19] but not all have agreed about the size of the Roman army.[20] If eight legions were present and normal troop strength is applicable, each legion would have contained about 5,000 men and likely was accompanied by as many Roman allies[21]—possibly fusing 8 Roman legions with 8 allied legions.

If legion strength was indeed normal and matched by allies, a number of 80,000 for infantry alone is most reasonable. The types of Roman forces would parallel mostly those detailed in the previous chapter: for infantry, an *acies triplex* of lightly armed and more inexperienced *hastati,* and heavier-armed and more-experienced *principes,* and *triarii* with the light infantry *alae* allies on the maniple wings.

The combined cavalry of two consular forces normally at 3,000 each made of equestrians and noble patricians was now doubled to 6,000. Perhaps the most telling for the huge Roman army is that due to the speed of levying and recruiting, perhaps half of the force were raw recruits with little to no training for a large portion of *hastati* up through many of the *principes* ranks. It is not unlikely that Hannibal knew this fact from his spies and prepared accordingly. The Roman primary offensive weapon was the short gladius sword or its predecessor, along with javelins and pilum spears, and defensively almost all had shields. The poorer *hastati* were more modestly protected than the wealthier *principes* and *triarii.* Unless individually wealthy, the Roman Allies would likely have been armed and protected more like the *hastati.* It is unknown but possible that the Roman velites were the legionary forces left in charge of the Roman camp.

Like at Trebbia River and Trasimene, Hannibal's army was also primarily a motley infantry force, mostly made up of mercenary veterans. At Cannae he had Carthaginians, Spanish Celtiberians, other "Africans" (mainly North African and often called Libyans), slingers from the Balearic Islands and Celts, some of whom were heavy infantry and some who made up a considerable portion of his allied cavalry. Hannibal's cavalry was primarily Numidian from North Africa. The number of Hannibal's infantry is debated but a general consensus suggests around 40,000 troops from these different cultural and language groups.[22] Most accounts suggest Hannibal was outnumbered 2:1, or 80,000 to 40,000 at Cannae.

While many of his multicultural soldiers were dressed in a bizarre array of styles and colors, as noted by Polybius, with naked Celts, Spanish Celtiberians in short tunics bordered with purple, altogether "strange and impressive," Hannibal also had many of his men hybridly armed with Roman weapons and armor collected from Trebbia River and Trasimene,[23] thus creating confusion, especially for the battlefield Romans. The primary swords of the Celts were long iron slashing swords with a greater reach than the presumed Roman *gladius*. Additionally, the Africans and Spanish Celtiberians would have had stabbing swords[24] and if the infamous steel *falcata* was present at Cannae as a Celtiberian weapon made by talented Spanish smiths known for their metalworking, it was a formidable one, capable of severing limbs in one blow, what some might also call a kopis.[25] Some *falcata* from this period exist in Spanish museums, and they look like brutal front-weighted cleavers with sword points.[26] Hannibal's war elephants were all gone, with none mentioned after the march to Trasimene in 217 BCE with but a single elephant noted in the progress through the Arno Marshes.

BATTLE CHRONOLOGY

Accepting Hannibal's provocations on the morning of August 2, against the advice of Paullus, Varro put out the red *vexillum* pennant on his tent post, the prepared signal for a day of battle.[27] Romans began crossing the river *en masse* to give battle where Hannibal's assembling forces faced north. Because it took hours to assemble such a large force, the morning was given over to lining up in formation on the Roman side facing south. A few skirmishes occured here and there as the Romans put out an advance light infantry of *hastati* and Balearic slingers and javeliners were sent out from Hannibal's camp to cover his deployment.

The Romans had little room between the Aufidus River and the hill ridge, but they lined up in this fashion. Varro placed himself on his left with the Latin cavalry and Paullus, submitting to Varro on his day of command, lined up on the Roman right side with the Roman cavalry near the river. The proconsul Servilius Geminus led the massed infantry center. The depth of the many hundreds of Roman infantry maniples must have reached nearly a mile, possibly even beyond the stream at first or angled back along the stream. However, they could only line up so wide due to the narrow plain constricted by the Aufidus stream, with the Allied forces on the flanks either between the cavalry and the rest of the infantry or in front of the rest.

Varro made a tactical mistake by not holding out around half of his force; by committing them all too early he left nothing in reserve except the 10,000 guarding the camp,[28] which would also have reduced battlefield compression, alleviating the massed Roman maniples since the narrow field of battle could only hold so many.

Hannibal placed his cavalry on his two wings as well, the Numidians led by Maharbal on his right facing Varro and his Latin cavalry, and the Celtic and other horse on his left led by Hasdrubal facing Paullus and his Roman cavalry. Hannibal took up the infantry center with the Celts and many of the Spanish, with the African infantry and possibly some Spanish infantry on the right and left sides hidden behind the cavalry. This hidden African contingent according to Livy would be hard to tell apart from Romans as they were armed with captured Roman shields and weapons,[29] likely a ploy of Hannibal. Given whatever space was between the two armies, Hannibal employed a fascinating formation in front, a gambler's tactic that would yield enormously by its trickery.

He had his center form a convex bulge, like a crescent moon that extended outwards. Polybius precisely records it in detail as a *mênoeides kúrtoma* ("crescent shaped bulge"). The libeccio wind from the south picked up by late morning and was already blowing back into Roman faces from the back of Hannibal's army. The dusty wind was possibly even a signal Hannibal may have waited for, since it was mostly an afternoon phenomenon of high pressure heating up over land and drawing sandy African air from the distant Mediterranean Sea to the southeast along the coast.

The battle began whenever the trumpets blew and the shouts rose over soldiers banging or thumping their shields. Immediately the Carthaginian cavalry under Hasdrubal charged to meet the Roman cavalry under Paullus, stirring up a cloud of dust, while the Numidian cavalry under Maharbal on the right only engaged the Allied Latin cavalry in skirmishes, dancing back and forth without engaging in full battle but stirring up more dust. Although it was so much larger and could have stretched twice as wide under normal circumstances to outflank Hannibal, the Roman infantry stretched no wider than Hannibal's front due to the topography. The left-side Carthaginian cavalry under Hasdrubal outnumbered the right side Roman cavalry and so forced them backward and northeast along the river, as Polybius describes "mercilessly cutting them down."[30] Here Paullus was already wounded and possibly out of commission, blood running down his face from a successful Balearic slinger projectile.

Was the wounding of Aemilius Paullus merely an accident or could Hannibal have directed the Balearic slingers to search for and single out Aemilius Paullus early on? He would possibly stand out on the Roman side by official dress and regalia and possibly be close to a legionary standard—especially if already on horseback—and these early light skirmishers were often closest to the enemy line before pulling back. Neutralizing the most effective Roman leader would yield Hannibal a huge early battle advantage.

When the front center of Hannibal's line met the heavy Roman infantry the ensuing crash of intense hand-to-hand combat and the ferocity that the Celts typically brought challenged the orderly Roman maniples, whose shorter swords had nowhere the reach of the Celtic long swords. But by sheer numbers the Romans pressing forward would expect to prevail, and Hannibal reinforced that impression by yielding backward, leaving his wings intact but bending his convex center into a concavity, making sure his heavy infantry center did not break against the inexorable Roman push. While some doubt this was a planned "retreat," Wheeler has noted such a stratagem as *phygomachein* and the mobile Numidians were also adept at it.[31]

Gradually the center's cavity seemed to deepen on the wings as the Carthaginian cavalry then chased the Roman cavalry out of sight right off the field. Up to this point Varro, whether he was aware of Paullus' plight or not, must have had some report or could see what looked like the Carthaginian center caving in. He could have heard the shouts of heavy Roman infantry who thought they were achieving victory in the middle of the battlefield despite the annoying wind blowing dust in their eyes. So eager to cave in the center, the surging Romans had mostly ignored their left and side wings or thought the wings were being filled with Roman comrades in the dust where visibility was murky, although it would turn out they were not Romans at all.

Hannibal was biding his time until every element of his battle plan was ready and the trap could be sprung. Regarding topography and Hannibal weaponizing nature, Appian also supports the battle factor of wind blowing dust and obscuring Roman vision (7.87.20).[32] Appian adds that Hannibal hid men and cavalry in gullies on a hillside, which may be the hill of the Cannae vicus itself under Hannibal's control on the south side of the battlefield. Whether the latter account is true or not—around four centuries after the fact—considerable subsequent erosion and alluviation over two millennia plus might make it difficult to know if these gullies could support hiding troops even though the hill does contain gullies that do not now appear to be so deep or significant.[33]

Suddenly the lulling Numidian cavalry raising dust on Hannibal's right at last charged Varro's Latin and Allied cavalry and they were immediately deadly in their attack. A surprised Varro fled the field; having no experience with the lethal Numidians on their wheeling horses who could dash in and out again dealing death before the Allied Romans could even regroup. Some seventy horsemen accompanied Varro in absolute haste, whether to protect their general or flee or both. Varro never returned to the battlefield but even if the Roman army was headless without Paullus—arguably better off without Varro—the other veteran officers were suddenly as surprised as their men because Hannibal revealed his ace card.

The deep pocket in the center changed where Romans were massed so tightly that only the outer edges of the pushing maniples were able to use their weapons suddenly changed. Instead of retreating the Punic forces held their ground. With perfect timing the hidden Africans made their move. They had first been concealed by Hannibal's cavalry on both sides and then only partly visible in the swirling dust stirred up by cavalry and the *libeccio* wind, a dust that blinded the Romans because they faced it, especially the infantry who were focused on their front, not their sides. Plus, Hannibal's Africans were armed so similarly to the Romans, as Livy noted, "one might have thought the Africans were Roman soldiers."[34] Now at a likely sounded signal—a high bugle playing just the right notes carrying over the din—Hannibal's Africans all turned inward together and began to batter the surprised Romans, employing pikes against the massed legions.

It quickly became a three-sided box made tighter as more Romans poured into it from behind, still unaware of the trap Hannibal had set. Then it became even worse for the Romans when very quickly Hannibal's heavy cavalry under Hasdrubal returned from chasing Roman horsemen and arrived behind the Romans. Possibly also alerted to move in by the same bugle or similar cue, they fully closed off the rear, making a full four-sided killing box from which there was no escape for the panic-stricken Romans, with so many of the new recruits never having faced death before.[35] The coordinated timing had to be just right for all to come together, including the returned cavalry, and by now it was mid-to-late afternoon. The killing, probably so much of it unnecessary, now began to be very one-sided.

Many commentaries over millennia have termed this appalling climax as Hannibal's double-envelopment maneuver (or treble depending on which phase). Modern historians also call it the "jaws of death" or the "vice"[36] and that certainly became the outcome as so many Romans were butchered—possibly 200 a minute[37]—as line after line of Roman soldiers were mown down in their maniples, officers, veterans, and recruits. If the Spanish falcata hacking away at the front was used, its heavy toll included dismemberment of limbs and maimed flesh.

Many Romans were likely rendered useless by being hamstrung where they stood, hardly able to crawl and nowhere to go except over other bodies. The battlefield screams and agonies were not Roman alone, because even though Romans died close to a ratio of 10:1 relative to their enemy, that still meant at least about 6,000 deaths in Hannibal's army, the brunt of suffering and toll worst among the Celts in the heaviest concentration at the battle's center. The

Roman defeat was probably as horrible to hear as to behold with the terrified young Roman adolescents weeping in shock, many dismembered and mired in their own blood and waste. Cannae was Rome's worst nightmare and if Hannibal was brutal he also had a point to make to all of Italy, not just to Rome by demonstrating his ability to completely neutralize Roman advantages.

When the afternoon carnage was over, Livy numbered at least 50,000 Romans who died that day at Cannae, whereas Polybius counts 70,000.[38] This discrepancy of 20,000 is curious but many historians tend to go with Livy's more conservative number, partly because so many thousands were captured who surrendered at the camp. Carthaginian total dead may have numbered just under 6,000.[39]

Is this discrepancy in the number of Roman dead between Polybius and Livy, a possible ploy on the part of Livy if Polybius was perhaps more credible? By possibly reducing already vast Roman mortality, was it an attempt by Livy to balance and equalize Cannae with the later Battle of Metaurus (207 BCE) where he records 56,000 Carthaginian fatalities there (Livy *Hist.* 27.49.6) to make the Roman number at

Cannae less politically horrid? We may never know the exact truth here about full Roman mortality at Cannae or resolve this discrepancy.

The battle took only about six hours total from the oblique morning light to the midafternoon shadows. At least a full legion or two survived including the 10,000 captured Romans left at the camp and the few—maybe seventy survivors out of 6,000 horsemen—who fled first to Canusium with Varro plus a few thousand other fleeing Roman survivors.

Four escaping tribunes, including the young Publius Scipio—the future Scipio Africanus—and the son of Fabius Maximus, went to Rome with Varro as quickly as possible. Livy says 19,000 Roman prisoners were taken at Cannae, which would amount to about two legions.[40] Subtracting 19,000 from 80,000 leaves 61,000, so Livy is vague as well. Some of the Roman cavalry may have escaped much of the destruction although Polybius may include many of these among the casualties accounting for his higher number of Roman dead.

O'Connell opines there were more dead soldiers that early August day at Cannae than any other single combat day in the entire history of Western battles.[41] Hanson suggests 30,000 gallons of blood were spilled at Cannae.[42] Birds of prey descended quickly on the field as most of the exhausted Carthaginians sat down to rest and others, human scavengers, began collecting battle trophies over the next few days. Hannibal must have joined the team of officers surveying and scouring the field where heaps of bodies lay, searching for his wounded and the dead Roman high command. Adding to the deaths of so many common Roman soldier-farmer citizens whose bodies lay crumpled at Cannae, a vast portion of Rome's experienced military leadership was also lost. Not only had the consul Aemilius Paullus perished but the other dead generals included Servilius Geminus and Marcus Minucius, the previous impetuous master of horse under Fabius Maximus who had also been a prior consul. Plus, 29 out of 48 tribunes were killed (more than half[43]), 80 senators dead (or elected magistrates of senatorial rank who would have served in the next Senate) and over 200 of *eques* rank perished, Roman knights whose gold rings were collected and counted on the field of death. No wonder the bloodshed at Cannae still makes us shudder.

UNUSUAL DECISIONS AND TACTICS

There was a concatenation of bad decisions by the Romans at Cannae set against the uncanny tactics of Hannibal. For the Romans, it was risky loading the army with so many raw recruits, possibly up to half or maybe more since Hannibal had depleted the legions twice before in the prior two years at Trebbia River and Trasimene. As mentioned, this new Roman army had possibly only been together training on the march for a week before the battle at Cannae. This was a liability Hannibal would be sure to exploit if he knew it, and it is not at all unlikely that he had this knowledge with his usual military intelligence gathering in preparation for battle. Too few historians have factored this as a part of the Roman disaster; Hanson certainly emphasizes it.[44]

Varro's decision to fight—blindly accepting Hannibal's offer without thinking it through—where the massed infantry had no room to maneuver between hill and stream would doom his army. He let Hannibal choose the terrain and it was stupid to do so because it would turn the primary Roman advantage of overwhelming numerical superiority into a disadvantage, exactly as Hannibal planned in turning a strength into a weakness and vice versa, because there could be no Roman outflanking of the enemy in the small space between the river and the hills.

Also a bad decision on the Roman side was the timing of Varro's flight from battle with his cavalry when his relief was most needed. The numerical cavalry advantage Hannibal had, almost 2:1, was enhanced by the fact that his Numidians were the best horsemen in the Mediterranean world with ample battle experience. Polybius notes that the cavalry advantage of Hannibal and by inference the cavalry deficiency of the Romans was a huge factor at Cannae, "contributing most to the victory of the Carthaginian's" as well as his suggestion that infantry should be reduced in favor of cavalry.[45] To have Numidians like Hannibal did was an advantage that young Scipio—who was evidently observing here—would later copy at Zama.

Hannibal's decision to again engage the Romans was first prompted by the need to provide ample food supply for his army in the coming winter; the grain supply near Canusium made that risk worth it. Knowing the Romans were en route to him with a massive army, Hannibal again chose the battleground, one that would advantage him and his mobile cavalry—one of his best assets—but box in the Romans who would crowd the battlefield with so much massed infantry. The coastal plain would likely give the Romans overconfidence in thinking they could outflank him but if he could choose the context before they arrived and use the river to his advantage, the narrowness of the battleground hemming in the Romans by the river could neutralize the plain's openness and the seeming advantage it gave to them.

Hannibal's choosing to fight on the day when Terentius Varro commanded or forcing his hand to respond first—especially when Aemilius Paullus was unhappy with the battleground and wanted to delay—was also another exploitation of the dual command predicament where the least experienced Roman consul would lead the field, especially in this case with one who was incompetent at war. The taking of the water source of the Aufidus before battle in the summer heat would dehydrate the Romans and then harassing any Romans who tried to get water would also serve to frustrate and anger Terentius Varro and make him eager to redress the matter. Some later accounts say Hannibal even polluted the water source in advance by throwing dead animal bodies to decompose in the Roman sectors downstream.[46]

Again using the environment as a weapon in his arsenal to compensate for his smaller army, Hannibal's positioning to the south with his army facing north was also apparently no accident; it was likely a deliberate stratagem where he chose to have the Romans face the heavy dust-bearing wind, the *libeccio* that blew summer grit from Africa into the Romans' faces and eyes, both an irritant and a likely impediment to their field vision.[47] This directional choice with the *libeccio* wind was even noted by more than a few ancient sources such as Appian, Plutrach, and Frontinus.[48]

Given the cramped space Hannibal selected, since the Romans could only fight on the outside of their maniples, (Hanson suggests, only 2,000–3,000 Romans out of 70,000 massed together could fight at one time[49]), much of the infantry would be so hard-pressed together their weapons would be virtually useless. In addition, Hannibal's African veterans would be facing the least experienced Romans on the wings of the maniples.

By luring the Romans into a trap with a pretended retreat of his center, after first employing his precisely-planned lunar crescent convex-shaped bulge (*mênoeides…kúrtoma* in Polybian Greek[50]) laden with Celts and Spanish Celtiberians at the front, Hannibal drew the Romans into a box with three sides, where his Africans and other Spanish Celtiberians, "the jaws," hammered the massed Romans from both left and right.[51] With his cavalry's return to the battlefield after chasing away the Roman cavalry, closing the box from the rear, the encirclement would be complete on four sides. As already stated, this double or multiple envelopment of his closing "jaws of death" is often said to be Hannibal's most brilliant move at Cannae.

Because Hannibal had such good military intelligence up to this point, he must have known the newness of Rome's recruits that filled so many ranks as stated above. He would also have known the psychological effect of panic and the dread in the pit of the stomach of the inexperienced legions, creating what is often termed mass psychosis in modern parlance, an effect that a general like Hannibal knew from decades of experience, mostly in controlling it rather than letting panic develop among his own troops.[52]

Given what we now know of war trauma and post-traumatic stress syndrome in war,[53] it should be no surprise if ancient military leaders of Hannibal's intellectual depth also knew how to wield such a factor to his advantage. Even his demeanor before battle getting his men to laugh with his Gisgo quip was a "psy-ops" type of tactic in building confidence or letting off the steam of nervousness before a battle. Centuries before Hannibal, Thucydides had also described just such battle stress (with its attendant "fear, grief, shame"[54]) and the literate Hannibal was deeply schooled in his youth in the history of the Greco-Persian and Peloponnesian Wars as well as in Alexander's tutelage by Aristotle.

Was Hannibal overconfident or was he just more aware of his enemy on both detailed micro (Terentius Varro's inclinations) and overarching macro (raw and inexperienced recruits) scale? It is most likely that Hannibal was already a master of the psychology of war at age 31. Hanson certainly attributes some of Hannibal's success at Cannae to the overwhelming panic of the demoralized young Romans who had no experience with military matters and were poorly led into these killing fields.[55]

How much of this step-by-step sequence at Cannae was the tactical shrewdness of Hannibal's original plan in his typical preparedness or was made possible by his equally brilliant extemporaneous adapting in the field can never be known. His retreat at Cannae may have been real; his box closing with returning cavalry possibly mere coincidence. He has to be given credit in choice of battlefield with its constricting topography. But judging by his similar enterprising tactics at Trebbia, Trasimene, Volturnus and now at Cannae, even though each required different plans—implementing his use of the environment, weaponizing nature by using the windborne dust in Roman eyes, exploiting his knowledge of the character of the enemy leadership against them—strongly suggests the Roman debacle and Hannibal's victory at Cannae was no mere accumulation of coincidences, stumbling into opportunity or mere good luck on Hannibal's part.

OUTCOMES

Hannibal's men captured the 10,000 Romans left to guard the camps and although up to 3,000 other Romans fled to Canusium, many of them cavalry officers like the escaping consul Terentius Varro—and the young tribune Scipio who would become Scipio Africanus,[56]—Rome once again, like at Trebbia and Trasimene, had faced a superior commander of genius. Beyond the almost incomprehensible loss of about 1 out of every 5 fighters reducing 20% of adult Roman males from their entire population on Cannae's battlefield, another fairly immediate outcome was a considerable defection of South Italians to Hannibal, bringing in what he had hoped for in both reinforcements and taking from a beleaguered Rome both territory and economic gain.[57]

Hannibal's Numidian commandant Maharbal urged him to march on Rome immediately, saying he could be feasting on the Capitolium in five days, less than a week of hard march if they started right away. But Hannibal knew this was impractical for his exhausted army and he could not muster enough horses to make such a journey of 250 miles, let alone besiege such a huge fortified city with many miles of walls, and because the small Celtiberian oppidum fortress of Saguntum had taken months of siege earlier in 219 BCE.[58] Maharbal famously quipped, "Hannibal, you know how to win but not how to take advantage of it."

As O'Connell develops in compelling narrative, the Roman "ghost" survivors of Cannae ultimately overcame their shame and were retrained in Sicily by Scipio, becoming the core of his Zama army fighting for redemption for the defeat at Cannae.[59]

Just as Livy previously reported about Trebbia but was now magnified exponentially:

> The news of Rome's defeat by Hannibal at Cannae triggered a severe crisis for the Roman State, as it plunged the government into fear and uncertainty, with a heightened expectation that Hannibal could himself appear before the walls of Rome, bringing direct violence to the very center of the Roman world.[60]

Livy describes the city filled with terror and confusion and the Senate meeting to defend the city, mistakenly thinking Hannibal was en route whereas he was now counting the quickly mounting defections from Rome in South Italy. Rome was not the only state that responded to its defeat with surprise, even Philip of Macedon far from Cannae was so impressed with Hannibal's victory there that he took up arms against Rome and threatened them on the southern Adriatic, a worrisome prospect for the Roman port of *Brundisium (Brindisi)*.[61] Now the Romans had another firefight besides Italy to quell.

CONCLUSIONS

Cannae has often been called the greatest—or most devastating—battle in history, where between 50,000 to 70,000 Romans perished in one afternoon. Yet Hannibal, who had never lost a full battle in Italy would go on to lose the war. Whether his overall doomed strategy can be faulted in trying to sustain an invasion for over 15 years and turn the rest of Italy against Rome, or whether the challenges he faced were ultimately insurmountable against a people who refused to throw in the towel and admit or accept defeat (unlike Hannibal's own home of Carthage) and who could raise legion after legion, or possibly for other concatenated reasons, the irony of Hannibal is that he could win great battles on tactical genius but not significantly change history by keeping Carthage independent.

Recalled to Africa, the Battle of Zama back in his Punic homeland in 202 BCE, where Scipio Africanus defeated him, was not necessarily Hannibal's low point nor likely even mostly his fault, but it did mark a sea change and the first vision of a Roman Mediterranean as a pre-cursor of eventual empire. Punic Carthage would soon be only a memory even as Hannibal's specter lived on. However ultimately futile in not achieving his highest goals, posterity recognizes Hannibal as one of history's greatest generals and certainly a tactical genius possibly without parallel.

As mentioned, there is no debate that Scipio learned tactics from Hannibal at Cannae and improved on them en route to subsequent victory at Zama. Polybius (*Hist.* III.117) as a former Greek cavalry officer noted Punic cavalry was decisive in Hannibal's victory—"superiority in cavalry wins battles"—and lamented Rome's lack therein, so future Roman victories like Scipio's at Zama involving more Numidians learned this lesson. Ultimately it may be asked how much Hannibal schooled Rome in war, and, however provocative an idea, whether Rome could have so easily and quickly conquered the surrounding world without learning from its defeats against Hannibal.

◆ ◆ ◆ ◆ ◆

ENDNOTES

1 Robert O'Connell, *The Ghosts of Cannae* (New York: Random House, 2011), 4.

2 Patrick Hunt, "Quintus Fabius Maximus Verrucosus," *Encyclopaedia Britannica*, 2014.

3 Ernle Bradford, *Hannibal* (Macmillan, 1981) London: Folio Society, 1998.

4 Frontinus, *Strategemata* I.5.28.

5 Paul Erdkamp, "Polybius, Livy and the Fabian Strategy." *Ancient Society* 23 (1992): 127–47.

6 Polybius, *The Histories*, III.107.9.

7 Michael P. Fronda, *Between Roman and Carthage: Southern Italy in the Second Punic War*. Cambridge University Press, 2010, 67–72.

8 Gregory Daly, *Cannae: The Experience of Battle in the Second Punic War*. London: Routledge, 2002, 16.

9 Dexter, Hoyos, "Cannae, Battle of." *Encyclopedia of Ancient History*. Malden, MA: Wiley-Blackell, 2015.

10 John F. Lazenby, *Hannibal's War: A Military History of the Second Punic War* (Warminster: Aris and Phillips, 1978), 77–79; Richard J. Talbert, A. *Atlas of Classical History* (London: Routledge, 1991 repr. ed.), 99.

11 Serge Lancel, *Hannibal*. Antonia Nevill, tr. (Oxford: Blackwell, 1998), fig. 6 a mile east, somewhat closer to the modern SS16 road.

12 Dexter Hoyos, "The Age of Overseas Expansion" in Peter Erdcamp, ed. *A Companion to the Roman Army*. Malden, MA: Wiley-Blackwell, 2010, 63–80, esp. "Cannae," 66–69.

13 Goldsworthy, 2000, 201.

14 Goldsworthy, 2000, 202.

15 Polybius, *The Histories*, III.110.8.

16 Polybius, *The Histories*, III.112.3–4.

17 Plutarch, *Life of Fabius Maximus* 14.3.

18 Plutarch, *Fabius* 15.2–3.

19 Polybius, *The Histories*, III.113.5; Goldsworthy, 2000, 200.

20 Stanley Arthur Cook et al., eds., *Cambridge Ancient History*, vol. VIII, Rome and The Mediterranean 218–133 BC, 1965.

21 Polybius, *The Histories*, III.107.10–12.

22 Nic Fields, *The Roman Army of the Punic Wars 264–146* BC. (Oxford: Osprey Publishing, 2007), 23; Lancel, 104, suggests nearly 50,000 with Hannibal.

23 Polybius, III.114.1–4.

24 Mark Healy, *Cannae 216 BC: Hannibal Smashes Rome's Army* (Oxford: Osprey Publishing, 1994, 2005 repr. ed.), 84.

25 *kopis* in Greek, Goldsworthy, 2000, 208. Healy, 2005 repr., 24, notes the *falcata* at Cannae, as does Richard Gabriel,

Scipio Africanus: Rome's Greatest General (Dulles, VA: Potomac Books, 2008), ch. 3.

26 e.g., Cerralbo Museum, Madrid, a 4th-3rd c, BCE artifact from the Celtiberian necropolis of Las Angosturas, Granada; another from Bastida de Les Alcusses, 4th c. BCE, Museo de Preistoria de Valencia; yet another falcata is modeled on a clay relief in the Etruscan Tomb of the Reliefs, 4th-3rd c. BCE, Cerveteri Necroplis, Italy. Dave Baker, a Hollywood weapon-maker replicated some iron-steel falcata and we tried them on a documentary set in 2010 where they easily severed animal bones.

27 Goldsworthy, 2000, 204.

28 Hanson, 2002, 107.

29 Livy, XXII.46.

30 Polybius, *The Histories*, III.115.4.

31 Everett Wheeler, *Stratagem and the Vocabulary of Military Trickery. Mnemosyne Supplement* 108. Leiden: E. J. Brill, 1988, a stratagem of "feigned retreat" or *prospoiete phyge*, "pretended flight."

32 Appian, *Roman History*, Book 7 (Hannibal). Brian McGing, ed. (Cambridge, MA: Harvard Loeb, 2019), 340–41 [*Appian* 7.87.20].

33 Hunt, Patrick. "Hannibal's Hard Homecoming after 203 BCE." *Acta Antiqua* 62.2–3 (2024) Academiae Scientiarum Hungaricae, 417–25.

34 Livy, XXII.46, Healy, 2005 repr., 84.

35 Fields, 38.

36 e.g., Healy, 205, "vice"; Fields, 38 "jaws"; Hanson, *Carnage and Culture*, 106, "Hannibal's Jaws."

37 Hanson, 2002, 104.

38 Polybius, *The Histories*, III.117.4, 70,000; Livy, XXII.49 & XXV.6, around 50,000. Several others, including Appian (*Hannibalic War* 4.25 "in those few hours") and Plutarch (*Life of Fabius* 16.8, "it is said 50,000 Romans fell in that battle") also round it out to 50,0000 but Quintilian (*Institutes* VIII.6.26) lists 60,000 dead at Cannae when talking of extending tropes. O' Connell, 3, counts 48,000 dead or dying Romans.

39 Polybius, III.117.6 says 4,000 Celts, 1,500 Spaniards and Africans and 200 cavalry; Lancel, 2000, 108. I might also be slightly tempted to side with Polybius here because Livy quotes the same amount of Carthaginians (55,000) perish later at Metaurus, and he may have tried to create propagandistic parity for justification of a Roman triumph equal to the Cannae loss.

40 Livy, XXII. 49. He numbers as Roman prisoners the 10,000 guarding the larger camp who had been kept out of battle, 7,000 who had fled to the smaller camp only to be captured and 2,000 who had fled and sought refuge in the fortified village of Cannae itself and were taken prisoner there.

41 O'Connell, 4.

42 Hanson, 2002, 104.

43 O' Connell, 160.

44 Hanson, 2002, 101, 107, esp. 108.

45 Polybius, *The Histories*, III.117.4.

46 Dio Cassius, *Fragments* 15.57.25.

47 Patrick Hunt, *Hannibal*. New York: Simon and Schuster, 2017, 120, 267–68

48 Plutarch, *Fabius* 16.1; Frontinus, *Strategemata* II.2.7.

49 Hanson, 2002, 108.

50 Polybius, *The Histories*, III.113.8; Fields, 38; Hanson 107.

51 Frontinus, *Strategemata* I.16.2–3, II.3.7.

52 Max Corvo, *The O.S.S. in Italy 1942–1945*. New York: Praeger, 1990, 7,144. in the theater of Italian operations for invading from Sicily to the Alps, Corvo and his O.S.S.–S.I. cohort used a compelling psychological warfare plan, plus careful intelligence gathering along with disinformation in order to facilitate the Allied invasion of Italy in several ways parallel to Hannibal's use of intelligence and counterintelligence.

53 Marc-Antoine Crocq and Louis Crocq, "From shell shock and war neuroses to posttraumatic stress disorder: a history of psychotraumology." *Dialogues in Clinical Neuroscience* 2.1 (2000): 47–55.

54 Thucydides, *History of the Peloponnesian War*, VII.75.

55 Hanson, 2002, 108.

56 Livy, *The War with Hannibal*, XXII.53.

57 Frontinus, *Stratgemata* I.17.2.

58 Goldsworthy, 2000, 215.

59 O'Connell, 4.

60 Gregory K. Golden, *Emergency Measures: Crisis and Response in the Roman Republic (from the Gallic Sack to the Tumultus of 43* BCE. Ph.D. Dissertation, Rutgers University, Classics, 2008, 16–7.

61 David Abulafia, *The Great Sea: A History of the Mediterranean* (Oxford University Press), 2011, 186.

BIBLIOGRAPHY

PRIMARY SOURCES

Appian. *Hannibalic War* 52–58.

_______. *Roman History*, Book 7 (Hannibal). Brian McGing, ed. Cambridge, MA: Harvard Loeb, 2019.

Dio Cassius. *Fragments.*

Frontinus, *Strategemata.*

Livy. *The War With Hannibal.* Aubrey de Selincourt, tr. New York: Penguin, 1965, 1983 repr.

Polybius. *The Histories*. Robin Waterfield, tr., and Brian McGing, notes. New York: Oxford University Press, 2010.

Plutarch. *Life of Fabius Maximus.*

Quintilian. *Institutes* VIII.6

Thucydides. *History of the Peloponnesian War.*

SECONDARY SOURCES

Bradford, Ernle. *Hannibal.* New York: Macmillan, 1981 and London: Folio Society, 1998.

Cook, Stanley Arthur, et al., eds., *Cambridge Ancient History, vol. VIII, Rome and The Mediterranean 218–133* BC. Cambridge: Cambridge University Press, 1965.

Corvo, Max. *The O.S.S. in Italy, 1942–1945*. New York: Praeger, 1990.

Crocq, Marc-Antoine and Louis Crocq. "From shell shock and war neuroses to posttraumatic stress disorder: a history of psychotraumology." *Dialogues in Clinical Neuroscience 2.1 (2000):47–55.*

Daly, Gregory. *Cannae: The Experience of Battle in the Second Punic War.* London: Routledge, 2002

Erdkamp, Paul. "Polybius, Livy and the Fabian Strategy." *Ancient Society* 23 (1992) 127–47.

Fields, Nic. *The Roman Army of the Punic Wars 264–146* BC. Oxford: Osprey Publishing, 2007.

Fronda, Michael P. *Between Roman and Carthage: Southern Italy in the Second Punic War*. Cambridge University Press, 2010,

Gabriel, Richard. *Scipio Africanus: Rome's Greatest General.* Dulles, VA: Potomac Books, 2008.

Golden, Gregory K. "Emergency Measures: Crisis and Response in the Roman Republic (From the Gallic Sack to the Tumultus of 43 BC)." Ph.D. dissertation, Classics, Rutgers University, 2008.

Goldsworthy, Adrian. *The Punic Wars*. London: Cassell, 2000.

______. *Cannae: Hannibal's Greatest Victory*. London: Cassell, 2001.

______. *The Complete Roman Army*. Thames & Hudson, 2011.

Hanson, Victor Davis. *Carnage and Culture: Landmark Battles in the Rise of Western Power.* Anchor Books, 2002.

______, ed. *Makers of Ancient Strategy: From the Persian Wars to the Fall of Rome*. Princeton University Press, 2010.

Healy, Mark. *Cannae 216* BC*: Hannibal Smashes Rome's Army*. Oxford: Osprey Publishing, 1994, 2005 repr. ed.

Hoyos, Dexter. "The Age of Overseas Expansion" in Peter Erdcamp, ed. *A Companion to the Roman Army*. Malden, MA: Wiley-Blackwell, 2010, 63–80, esp. Cannae, 66–69.

______. "Cannae, Battle of." *Encyclopedia of Ancient History*. Malden, MA: Wiley-Blackell, 2015.

Hunt, Patrick. *Hannibal*. New York: Simon and Schuster, 2017, ch. 15, "Cannae", 131–47, 294–97.

______. "Hannibal's Hard Homecoming after 203 BCE." *Acta Antiqua* 62.2–3 (2024) Academiae Scientiarum Hungaricae, 417–25.

______. "Quintus Fabius Maximus Verrucosus." *Encyclopaedia Britannica*, 2014.

Lancel, Serge. *Hannibal*. Antonia Nevill, tr. Oxford: Blackwell, 1998 (originally Librarie Arthème Fayard, 1995).

Lazenby, John F. *Hannibal's War: A Military History of the Second Punic War*. Warminster: Aris and Phillips, 1978.

O'Connell, Robert. *The Ghosts of Cannae*. New York: Random House, 2011.

Talbert, Richard J. A. *Atlas of Classical History*. London: Routledge, 1991 repr.

Wheeler, Everett L. *Stratagem and the Vocabulary of Military Trickery*. Mnemosyne Supplement 108. Leiden: E. J. Brill, 1988.

ABOUT THE AUTHOR

Patrick Hunt (Ph.D., Institute of Archaeology, UCL, University of London, 1991) is an archaeologist and historian who has taught at Stanford University since 1993. National Geographic Society's Expedition Council sponsored his Hannibal Expedition in 2007–08. He was Director of the Stanford Alpine Archaeology Project from 1994 to 2012 and was elected as a Fellow of the Royal Geographical Society (London) in 1989. He is also an elected Fellow of the Explorers Club New York. Hunt is also a National Geographic Explorer since 2007 and a National Lecturer for the Archaeological Institute of America, including the prestigious AIA Norton Lectureship 2022–24.

Dr. Hunt has published more than 100 peer-reviewed articles on such diverse topics as monuments like the Pantheon, ancient notables such as Gyges from Herodotus, linguistics, biblical studies, the origin of Byzantine Silk, studies in Hebrew poetry and literary wordplay, Roman monuments in operas, calendrical megaliths, Olmec and Maya sculpture, iconography on Greek vases, myth palindromes, nautical exploration, art history, Egyptian stone working, Phoenician lore, and geoarchaeology among many other topics.

Dr. Hunt's articles on a range of topics, including Hannibal and Carthage, can be found in *Encyclopædia Britannica, Journal of Roman Archaeology, Studia Phoenicia, World Archaeology,* and Wiley-Blackwell's *Encyclopedia of Ancient History,* among others. He has authored more than two dozen books including *Hannibal* (2017), *When Empires Clash* (2015), *Alpine Archaeology* (2007), and *Ten Discoveries That Rewrote History* (2007).

He has been a featured scholar of the National Geographic Explorer TV program "Death of the Ice Man" and National Geographic / PBS / NOVA "Ice Man Murder Mystery" as well as for multiple episodes for archaeology and history on the History Channel.

He has lived in London, Athens and Jerusalem as well as annual time spent in Switzerland, France, and Italy every year since 1994, among many other countries, and has also conducted archaeological research in Peru on Inca sites and on Olmec, Maya, and Aztec cultures in Central America. Hunt has also been a postdoctoral Research Fellow in the Department of Near Eastern Studies, University of California, Berkeley, under Prof. David Stronach (1992–94).